D1314327

WE CAN READ about NATURE!™

BUSY BUILDERS

by MELISSA McDANIEL

BENCHMARK BOOKS

MARSHALL CAVENDISH
NEW YORK

With thanks to
Susan Jefferson, first grade teacher at Miamitown
Elementary, Ohio, for sharing her innovative teaching
techniques in the Fun with Phonics section.

Benchmark Books
Marshall Cavendish Corporation
99 White Plains Road
Tarrytown, New York 10591
Website: www.marshallcavendish.com

Photo Research by Candlepants, Inc.

Cover Photo: *Photo Researchers, Inc.*, Kenneth H. Thomas

The photographs in this book are used by permission and through the courtesy of: *Photo Researchers, Inc.*: Alvin E. Staffan, 4; Dan Guravich, 5, 13; Mark Boulton, 6; Wayne Lawler, 7, 14 (bottom), 26; Ken M. Highfill, 9; Francois Gohier, 11; Ron Austing, 12; Tom & Pat Leeson, 14 (top), 20, 21; R. Dev, 15 (top); Craig K. Lorenz, 15 (bottom); Kenneth H. Thomas, 16 (top); R.J. Erwin, 16 (bottom); John Mitchell, 17 (top); Gary Rethrford, 17 (bottom); Stephen J. Krasemann, 18, 22; Dr. Paul A. Zahl, 19; Jim Steinberg, 23; Jeff Greenberg, 24; Tom McHugh, 25; Kjell B. Sandved, 27 (top); Mitch Reardon, 27 (bottom); Gregory K. Scott, 28; David Frazier, 29. *Peter Arnold:* BIOS (M.Gunther), 8. *Animals Animals:* Doug Wechsler, 10.

Library of Congress Cataloging-in-Publication Data

McDaniel, Melissa.
Busy builders / Melissa McDaniel.
p. cm. – (We can read about nature)
ISBN 0-7614-1255-7
1. Animals—Habitation—Juvenile literature. [1. Animals—Habitation.] I. Title. II. Series.

QL756 .M355 2001   591.56'4—dc21   2001025431

Printed in Italy

1 3 5 6 4 2

Look for us inside this book.

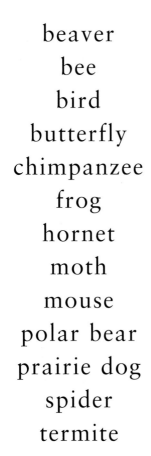

beaver
bee
bird
butterfly
chimpanzee
frog
hornet
moth
mouse
polar bear
prairie dog
spider
termite

Who builds houses?
Not just people.
From little birds . . .

A warbler

to big bears,

Polar bears

animals are builders too.

Some of the smallest animals
build the largest homes.
Who made this huge mound?

Tiny termites!

Chimps build beds high
in the trees.

Prairie dogs dig tunnels
deep underground.

Even some frogs build.

A tree frog

They make mud rings by the
water's edge.
Their babies grow up safe inside
where fish cannot eat them.

Birds fly near and far to find material . . .

An oriole with young

A pair of bald eagles

to build nests for their babies.

Some use twigs.

A heron pair

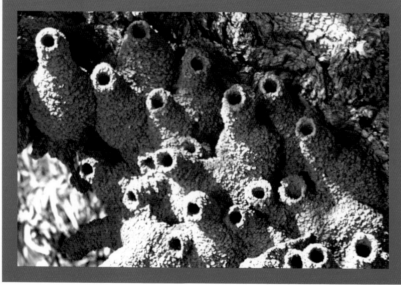

Some use mud.

Mud nests of fairy martins

Some
weave
grass.

*A weaverbird
at nest.*

Some
peck
holes.

A Gila woodpecker

Bugs build with many things.
A bee makes her nest of wax.

A hornet
builds her
nest from
paperlike
material.

16

A caterpillar spins a silk cocoon.
Inside, it changes to a moth
or butterfly.

A cecropia moth

Monarch butterflies

17

Spiders build traps to catch food. Some spin sticky webs.

An orb spider

Some make holes with lids on top.
They pop out and grab insects.

A trapdoor spider

Beavers are master builders.
They use their sharp teeth
to cut down trees.

Then they roll the logs to the water.

Beavers build long dams to make ponds.

In the ponds, they build
cozy houses.

In the city . . .

A stork

and the country . . .

A harvest mouse

in fields . . .

Termite mounds

Weaver ants building a nest

Weaverbirds nests

and in forests . . .

27

animals are busy builders.

A cliff swallow

Are you a builder too?

fun with phonics

How do we become fluent readers? We interpret, or decode, the written word. Knowledge of phonics—the rules and patterns for pronouncing letters—is essential. When we come upon a word we cannot figure out by any other strategy, we need to sound out that word.

Here are some very effective tools to help early readers along their way. Use the "add-on" technique to sound out unknown words. Simply add one sound at a time, always pronouncing previous sounds. For instance, to sound out the word **cat**, first say **c**, then **c-a**, then **c-a-t**, and finally the entire word **cat**. Reading "chunks" of letters is another important skill. These are patterns of two or more letters that make one sound.

Words from this book appear below. The markings are clues to help children master phonics rules and patterns. All consonant sounds are circled. Single vowels are either long –, short ˘, or silent /. Have fun with phonics, and a fluent reader will emerge.

The suffix "est" combines the short "e" sound with the consonant cluster "st."

s m a l l e s t   l aR r g e s t   f o r e s t   n e s t

The letter "g" will make the /j/ sound when the "g" is followed by the letter "e."

l aR r g e s t   c h ā n g e s   h ū g e

If the "y" at the end of a word is the only vowel in the word, the "y" will sound like "i."

b y ī   f l y ī

If there is a "y" at the end of a word and there is another vowel in the word, the "y" makes the long "e" sound.

t ī n y ē   s t ĭ c k y ē   S c ĭ t y ē   c ō z y ē

fun facts

- Each termite mound is home to millions of termites. The mounds are as hard as cement and can last hundreds of years.
- Spider silk is one of nature's strongest materials. It is even stronger than steel.
- Weaverbirds live in large groups. Some trees have two thousand weaverbird nests in them and have even lost branches because of the weight of the nests.
- Chimpanzees build a new bed every night. It takes about five minutes.

glossary/index

about the author

Melissa McDaniel is a writer and editor living in New York City. The author of more than a dozen books for young people, she has written on topics ranging from movies to ducks. She is an avid hiker and loves wandering through New York City parks, marveling at all the wild creatures that live there.